BLUETOOTH SURVIVAL GUIDE

FOR LAWYERS ™

By
Harold L. Hingston

H. L. Hingston & Company at this time is a start up small publishing company and is in the process of implementing a web site www hlhingston dot com which will include electronic e-mail facilities and a customer service section. If you are the purchaser of this book H. L. Hingston & Company intends to do whatever it takes to insure that you are a happy customer. It is the intention of H. L. Hingston & Company to build its business based on repeat business and not by trying to make a quick dollar out of new customers.

Good luck with this book and have a nice day.

H. L. Hingston & Company.

Table of Contents

PREFACE

This book is a first attempt to publish a series of books electronically. I had spent several years operating a patent application service on the Internet. However during the present severe economic recession or depression I found myself in a situation where I had no income. Of course, this was due to lack of business. Patents are expensive and during the present severe recession or depression people simply could not afford to pay for normal patent application services. So, I found myself with time on my hand's and at the same time an interest in the subject of Bluetooth applications for a small network.

I had the necessary digital hardware and software and time on my hands so I decided to start up a book publishing company on the internet. The first book is a user guide for the average person without a technical background. The information presented in this book is written in language that is aimed at the average person. As a trial lawyer my training and experience would clearly qualify me for this task because a trial lawyer is trained to take complex trial issues and describe them in simple language that the average juror can understand.

In order to put together the needed facts about Bluetooth devices I began reading the literature on the subject and even installing some Bluetooth devices. With my patent application service more or less dormant I found myself with four computers that were idle. Two computers are laptop devices and two are desktop devices. My original plan or goal was to set up a small area network using Bluetooth adapters on all of the computers. I later learned this is called a personal area network (PAN). I spent upwards of six months installing and reinstalling Bluetooth devices, testing the devices, and reading literature and documentation on the subject of Bluetooth. The subject of Bluetooth seemed to be a bottomless pit. Specifications for Bluetooth devices alone takes up hundreds of pages and all of these specifications are good for is to help an electrical engineer or computer scientist who is in the process of implementing Bluetooth hardware and software devices for sale to the general public.

As an out of work patent lawyer I had no interest in designing Bluetooth devices. My interest at this time it is to learn how to install and use Bluetooth devices in my patent application business once it picks up. This was about six months ago and since then business has not picked up in all. In fact it seemed that is becoming more and more difficult just to survive in our deep recession or depression at this time. With nothing else to do I decided to use my computer scientist education and experience to write a book that others could use to install and use Bluetooth devices.

In order to understand Bluetooth devices it is necessary to first understand the present devices that are used in networking. Networking with digital devices can involve as few as two digital computers or it can involve the entire world wide web or Internet. Fortunately the hardware and the software to implement any size of a computer to computer network is basically the same for either wired networks or so-called wi-fi networks or wireless networks.

However, Bluetooth devices are only used in networks between digital devices and at the present time only use very short range wireless technology. Wired and wireless or so-called wi-fi networks both use the basic transport technology over ethernet. Bluetooth devices use completely different transport technology, completely different hardware, and even completely different software.

All computers have a so-called operating system. The operating system of a computer manages the structure of the file system, manages the allocation of memory including so-called fast memory or random access memory, manages the operation of external devices such as a printer, manages input and output to the central processing unit, and manages the operation of so-called computer applications such as word processors, databases, spreadsheets, and other applications. The Bluetooth system unlike traditional network systems which are operated by the operating system of the computer, are basically a complete operating system themselves and operate the Bluetooth hardware as a sub – operating system of the basic operating system of the digital device or computer. This means that to describe how Bluetooth devices work, how to purchase them, how to install them, and how to operate them a completely different approach has to be taken than the usual approach of simply describing hardware and software that is operated by the computer operating system.

The purpose of this book is to describe how Bluetooth hardware devices and software can be used to save the user both a considerable amount of time and a considerable amount of money. This book is not a course in Bluetooth technology for the electrical engineer or computer scientist. Rather the objective of this book is to describe to the average person without a technical education how to use Bluetooth devices and Bluetooth software. The book is targeted to the market of practicing lawyers and their families. This is because this market is expected to be able to benefit from consumer oriented Bluetooth technology more so than almost any other group of people in our society. For example, to save a lawyer 10 hours of his or her time using Bluetooth technology as is described in this book could save the lawyer upwards of $1500 in billable time. Also to consider is the saving of aggravation wasted time wasted money and frustration for a person without an electrical engineering or computer science education or experience. Remember this book is the outcome of upwards of six months of study of Bluetooth related technology and in addition to that another six months of preparation of this book into a format that hopefully can be readily understood by the average person. Since lawyers have a reputation of being relatively intelligent with relatively high intelligent quotients it would seem that this book should be able to benefit lawyers as well as other persons.

If this book does nothing else it is my sincere hope that any reader of this book who drives a car and who uses a so-called mobile phone will use the information in this book to stay out of jail. It is a big problem for thousands of people in our society to be able to operate their car with both hands on the wheel and still carry on conversations or text messaging with their mobile phone. This is a problem because it is a criminal offense to drive a car and at the same time use a mobile phone to send text messages or even voice messages and not be able to drive a car with both hands on the wheel. To solve this problem all a person has to do is to read this book and install a Bluetooth headset on his or her mobile phone and use the headset for voice and possibly even text messages. This will result in the person obeying the law by driving the car with both hands on the wheel and at the same time using his or her mobile phone. It is believed that any mobile phone that is on the market today will support a Bluetooth headset. However there are always exceptions to every rule so you are advised to read this book first before you purchase any Bluetooth equipment or any Bluetooth headset to use with your mobile phone. For your information, a Bluetooth headset that will enable you to use your mobile phone hands free can be purchased for as little as $20.

This book can be used as a install guide, a user guide, a buying guide, and as a help with investing the in high technology Bluetooth related companies. This book is intended to be only a help with any of these projects. The book is not designed to replace the services of a computer programmer. Nor is the book designed to eliminate help that a person might offer at an electronics store to a prospective buyer of a Bluetooth hardware or software device. The book is intended to set out enough information about basic Bluetooth hardware and software technology so that a person can ask intelligent questions about the subjects. I have found that a person never has too much information about Bluetooth technology when the person is attempting to implement a Bluetooth related project. Hopefully this book will present enough information for the average person to make intelligent decisions about Bluetooth hardware and Bluetooth software technology without having to buy additional books to cover basic gaps in understanding.

The book is divided into seven chapters and two appendixes. Appendix A. sets forth some Bluetooth words and phrases with a brief description of their meaning. This appendix can be used as a reference on time to time. Appendix B. lists several additional sources for Bluetooth technology information. However, the only additional Bluetooth information that I have been able to find seems to be only useful to an electrical engineer or computer scientist who is in the process of designing Bluetooth hardware or software.

Chapter 1 – BLUETOOTH IN ACTION describes how Bluetooth technology can be used in day to day situations to save a considerable amount of time and/or money.

Chapter 2 – BASIC NETWORKING describes how traditional computer to computer networks are connected physically.

Chapter 3 – BASIC AREA NETWORKING (LAN) describes how traditional computer networks are connected to each other with software that transports graphics and data and software that is used to process the transmitted graphics and data like a basic word processor would do.

Chapter 4 – BLUETOOTH NETWORKS (PAN) describes how Bluetooth devices are connected physically.

Chapter 5 – BLUETOOTH DEVICES describes how Bluetooth devices have been connected and used with each other in real life and how a person can also connect and use Bluetooth hardware and software to connect and use computers and their peripheral devices.

Chapter 6 – PROJECTS describes situations involving the confusing Bluetooth technology situation in the world of today and how a person can at least try to avoid major pitfalls in using Bluetooth technology.

Chapter 7 – TRENDS describes how Bluetooth technology is now being used and developed behind the scenes so to speak by companies in our society and how some of these companies might stand to benefit financially in the immediate and long-term future using Bluetooth related technology.

APPENDIX A – BLUETOOTH BUZZ WORDS AND PHRASES lists some of the terms used by Bluetooth technology that a person might need to understand in order to make intelligent decisions about a Bluetooth project.

APPENDIX B – BLUETOOTH REFERENCE SOURCES lists some sources for Bluetooth information. Some of this information might be confusing or not understandable by the average person without a technical education.

Please keep in mind that Bluetooth technology is a breakthrough technology. Not only that but Bluetooth technology is in the development stage as of today. This means manufacturers might change hardware and software for a Bluetooth device without being able to update hardware or software installation instructions for the device. Even specifications printed for a Bluetooth device can be out of date. This book cannot possibly even try to keep up with all of this or with each and every development in Bluetooth technology. What is presented in this book is basic information about Bluetooth technology and practical usage that the reader can use to make intelligent decisions about any Bluetooth related project.

Parts of this book may seem dry and confusing to some readers. This is understandable but the more a reader understands this entire book the better he or she is to deal with the sometimes complex issues of a Bluetooth project. The key to understanding the mystery of Bluetooth is to understand that Bluetooth technology is in a world of its own and has a language of its own. A dictionary to this language is Appendix A of this book. Use this dictionary and it will be a big help to you in understanding any Bluetooth project. At least that is the intention of the author of this book.

With all of this said it is hoped that this book can be used by you for many months and years and that the information in this book can be used by you, by members of your family, and by your friends to stay out of jail and to save a considerable amount of time and money using Bluetooth technology.

Harold L. Hingston

CHAPTER 1 BLUETOOTH IN ACTION

DAY IN THE LIFE OF JOHN.

Today, John an attorney woke up in bed before the radio alarm went off. Mary and the kids were still asleep and he didn't want to wake them up this early in the day. So he hooked up his Bluetooth headset to his ear and his smart phone so they wouldn't hear anything. He checked his electronic e-mail, his text messages, the stock market and the price of his IBM stock from his smart phone. No one else heard any of this because the messages he got went directly to his Bluetooth headset hooked to his ear and the volume was so low it didn't disturb anyone else.

John shaved and took a shower then got dressed for work. While he was eating his breakfast he checked his electronic e-mail again using his Bluetooth headset and his smart phone. He could do this while eating because the Bluetooth headset allowed him to use his smart phone hands free.

After a quick breakfast John again hooked up his Bluetooth headset to his smart phone so he could drive the car hands free of the phone. This was a lucky day for John. On the way to work his car was pulled over by a police officer at a roadblock. The officer asked John to show him his cellular phone and John pulled it out of his pocket and showed it to the officer. The officer told him he could continue on his way to work because it was evident that

John had both of his hands on the wheel of the car so he was obeying the law. When John left the roadblock he noticed several other cars that were pulled over to the side of the road with police writing tickets. He assumed that these tickets were for drivers who were using their cellular phone with their hands or even using text messages while driving which is against the law in his city.

On this day John had to be in court for a hearing before going to his office. To be sure that his client would be in court for the hearing John called him using his Bluetooth headset connected to the smart phone and arranged to meet the client before the hearing to discuss what was expected to take place at the hearing. John left his smart phone on and connected to his Bluetooth headset so he could take any calls without disturbing anyone else before he and the client got into the courtroom for the hearing.

After the hearing John stopped at a local restaurant for coffee and a bagel then paid the bill with his Bluetooth enabled smart phone. The bill was paid electronically and the amount was automatically deducted from his bank account over the internet connection of the smart phone. He did this because his Bluetooth enabled smart phone automatically connected to a computer at the restaurant that accepted his electronic payment of his bill. This automatic connection is called an ad hoc network connection which is one of the network connections used by Bluetooth enabled digital devices.

After he arrived at the office John connected his Bluetooth headset and smart phone to the office Bluetooth intercom so he could communicate using his Bluetooth hands free headset with the receptionist and his associates. He also connected his laptop computer he used at his desk to the office wireless Bluetooth Personal Area Network (PAN) so he could use the office printer and facsimile machines directly from his laptop if needed.

After updating his time sheets, billing information, and other data from his desk using the laptop he transferred the updated data to a database on one of the office PC computers using the wireless Bluetooth Personal Area Network (PAN).

After work John stopped at the bank to get some cash from the bank ATM. His Bluetooth enabled smart phone made an automatic ad hoc network connection to the ATM and he used the keyboard on the smart phone to complete the cash withdrawal. The ATM printed a receipt of the transaction and John was back in business.

On the way home from work he called his wife and kids using the hands free headset connected to his Bluetooth enabled mobile smart phone. After John got home he disconnected his Bluetooth connections and drank a couple of beers while he watched a baseball game on television to unwind from his work day.

During the day one of John's contacts gave John his business card which John stored on his small handheld computer. John forgot about the electronic business card until he got into the connection range of the family PC. The PC computer made an automatic ad hoc Bluetooth network connection to John's hand held computer and transferred the electronic business card to the PC computer for permanent storage. The ad hoc Bluetooth network connection was then closed before John even realized what was happening. This electronic business card transfer happened before John had disconnected his Bluetooth enabled hand held computer.

DAY IN THE LIFE OF MARY

After Mary got out of bed and prepared several kinds of breakfasts for the individuals in the family she took a shower and got dressed for her new day. One of her first tasks for the day was to prepare a grocery list and shopping budget on the family PC computer. After the list was done she transferred it to her small hand held computer using a Bluetooth Personal Area Network (PAN) connection. She hooked up her Bluetooth headset to her smart phone and checked the news for the day. She also checked her electronic e-mail and her text message folder on her Bluetooth enabled smart phone, hands free.

On her way to the grocery store she stopped for gas for the car and paid for the gas at the pump using a Bluetooth ad hoc network and her Bluetooth enabled smart phone. The cost of the gas was electronically deducted from the family bank account. The security features of the Bluetooth connection gave her added protection against identity theft compared to the use of a physical credit or debit card at the gas pump. In fact the added security features of the Bluetooth connection are so secure that she considered her transaction to be invulnerable to anyone trying to steal her identity at the gas pump.

The family had recently suffered the tragic loss of their pet watchdog who just recently had passed away. While at the gas station Mary got an idea of how she might replace the family pet. This idea came to her unexpectedly as the result of well known feminine intuition. Her idea is to replace the departed family pet with a parrot at the bird shop and have this be a surprise to everyone else in the family. So Mary decided to drive to the bird shop to check out the availability of a pet parrot for the family.

Her car was a recent model equipped with a Bluetooth enabled so called hands free console. This let Mary use voice command technology to ask the smart phone connection to provide her with the address of a nearby bird store. An address was provided and she again used the hands free car Bluetooth console to use the global positioning system (GPS) system in the car to give her directions to the store from her present location. After getting the needed directions Mary drove over to the bird store and checked out several of the pet parrots. She particularly liked one parrot and its mate and she took photos of the two love birds with her Bluetooth enabled camera to show to the other members of the family before she actually bought the two birds.

Mary then continued her journey to the grocery store and bought the items on her shopping list and totaled the cost to be sure she was within her food budget using her hand held computer. All was well so Mary proceeded to checkout at the grocery store and paid her grocery bill using the ad hoc network feature of her Bluetooth enabled smart phone. She knew that the added security features of the Bluetooth connection would protect her against any attempt at identity theft at the cash register.

Her shopping done Mary headed for home in the car while continuing to use the Bluetooth headset connected to her Bluetooth enabled smart phone. She had several phone conversations while driving the car with both hands on the wheel as required by law.

When she got home Mary wirelessly transferred her photos from her Bluetooth enabled camera to the Bluetooth enabled photo printer and printed out the pictures directly from Bluetooth enabled camera to the Bluetooth enabled printer

Later all of the other members of the family agreed to get the two parrot love birds as pets.

DAY IN THE LIFE OF JOHN, JR AND BABY SISTER

It was a school day for the two kids and nothing unusual happened. Both of the kids had small hand held computers that they used to make notes for homework and to send and receive electronic e-mail and text messages. When they got home they made a Bluetooth connection to the Bluetooth enabled printer and made paper copies of their homework assignments and notes. Baby sister transferred a photo of the two love birds Mary had stored on her digital Bluetooth enabled camera to her hand held computer using the graphics file transfer feature of her Bluetooth connection. She then sent the photo to one of her girl friends to get her ideas about having parrots for pets using her smart phone.

Hopefully this will illustrate the possibilities of using each one of the three networking features of a Bluetooth connection and how this breakthrough technology can save a person both a considerable amount of both time and money.

CHAPTER 2 BASIC NETWORKING

WIRED NETWORKS

Any two digital computing devices that can be joined together can consist of a Local Area Network (LAN) although most LAN networking involves more than two devices. It is possible to connect devices using either serial or parallel wired connections. Serial line connections transfer one digital pulse or bit at a time whereas parallel line connections transmit bits in multiple of 8 bits at a time and therefore are much faster at data and graphics transfer than the serial line connections. However, recent developments in serial line connections have resulted in extremely fast data and graphics transfer so for all practical purposes serial line transfers are fast enough for LANs.

At present serial line connections between computing devices are limited to cable connections between small hand held computers and a PC computer or specialized computing or printing devices linked to a computer such as a label printer or a photo printer. In order for a serial line connection to work properly there must be an operating program on all connected devices that can operate both of the connections of the devices. If the operating programs are installed on the devices to be connected with a serial line connection all that is necessary to establish the linking process is to simply attach the appropriate serial line cable to each of the devices to be linked.

Most LAN networks involve several PC and/or laptop type computers connected together. This enables the linked computers to share devices such as a scanner, printer, and facsimile machine resulting in a significant savings in hardware costs. Most LAN wired connections involve installing an ethernet card in each computer to be connected and then using a cable called RJ-45 standard cables. This RJ-45 standard is established by the Institute for Electronic and Electric Engineers (IEEE) and must be complied with by any manufacturer of RJ-45 cables.

In order to connect two computers together in a LAN a hardware device called a switch is used. Each computer is connected to the switch and when one computer is linked to the other the link connection is switched from one computer to the other. The basic idea is the same when several computers are connected to the switch. When one computer sends a request to the switch to link to another specific computer connected to the switch the switching device establishes the requested link to the other computer. A switch is a device similar to what is called a network hub but more advanced in operation than a hub.

When a computer on the LAN or standing alone is connected to another type of device such as a printer using an ethernet wired connection the printer or other type of device is simply connected to the switch as if it were another computer on the LAN. When a computer requests a connection to the printer or other type of device the switch establishes the desired link. This way each computer on the LAN can connect directly to the LAN enabled printer or other type of device instead of sharing a printer or other type of device with another computer to which the printer or other type of device is connected. Not all printers or other devices can connect directly to a wired LAN instead of having to connect to one of the LAN connected computers then shared with other computers on the LAN. The printer or other type of device must be LAN ethernet connection enabled for this to work. Also each computer that uses the hardware printer must individually have the printer operating program called the printer driver installed.

A LAN can be connected to another LAN or to the internet by a device called a router. A router is similar to a switch or hub except that it connects two or more networks together instead of two or more computers. When a person establishes a connection to an internet service provider (ISP) connected directly to the web using a telephone, television cable connection or digital subscriber connection (DSL) the ISP provider usually furnishes the customer with a router used to connect LAN computers to the internet using RJ-45 cables. A customer of an ISP can even use a router to connect to the internet then re-connect to a LAN which is linked with a switch or hub so only one router is needed to connect one or more networked computers to the internet using the usual ISP approach.

A wired LAN computer can directly connect to some devices that are ethernet enabled machines. Such machines include a LAN enabled printer, scanner, and even a facsimile machine either individually or in a so called all in one printer, scanner, and facsimile machine. However, unless the machine or machines are ethernet enabled for direct connections the machine or machines need to first connect to a computer on the LAN then usage of the machine or machines is shared on the LAN by the connected computer. For a computer to use a direct connection to a printer or all in one type of printer, scanner, and facsimile machine each computer individually must have the operating software for the machine called the device driver installed.

For a wireless LAN setup there are generally two modes of operation   What is called peer-to-peer (PPP) serial connections is where all computers or wireless enabled devices such as a printer are interconnected. This is identical to when wired LAN computers and LAN enabled devices are interconnected by a switch or hub device. Another type of operation is called a wireless access point connection. This is identical to a wired LAN using a connection to a router with the router connecting the LAN devices and also another network the internet. In other words a network to network wireless connection. Each computer needs a wireless network card or has built in wi-fi capability to connect to the wireless LAN. If the wireless LAN uses a so called access point mode of operation then the wireless access point is a separate device usually connected to the internet and also to the wireless LAN.

Some printers or other machines can connect directly to each wireless LAN computer if the printer or other machine is so called wi-fi enabled. A wireless LAN is especially useful when one or more laptop computers are connected with one or more PC computers and one or more printers or multi function printer devices. A so called wireless LAN is similar to a Bluetooth personal area network (PAN)

---

except the PAN is more versitile and can cost a lot less to implement.

The object of this chapter is to identify many of the features of various kinds of LAN configurations. Many details are omitted in order to simplify the situation and at the same time provide a basis for understanding how Bluetooth networks work. Bluetooth enabled devices now in use or under development can replace either of the wired or wireless network configurations described in this chapter either in whole or in part.

The next step in wireless networking for digital devices is the so called AX-25 connection using long range radio transmitters and receivers instead of the wired internet. This AX-25 connection has been in use already for many years using the Linux operating system and amateur radio equipment. This wireless internet is often used by foreign agents or so called spies in the United States when communicating with their foreign associates. Of Course, Big Brother has a large radio transmitter-receiver in Washington, DC, to attempt to monitor this type of communication. As communication equipment continues to be built more and more on micro-technology or even so called nano-technology it is expected this long range radio network will perhaps someday replace the present wired internet with its so called domain name service. (DNS) This, however, remains to be seen at this time. Believe it or not Big Brother is even attempting to develop a communication system beyond that of the so called electro-magnetic spectrum so even the AX-25 radio network is subject to becoming obsolete someday

CHAPTER 3 LOCAL AREA NETWORKS (LAN)

BASIC LAN SOFTWARE

Bluetooth technology has been under development now for over ten years but has thus far been rather slow to replace existing digital networking technology. Probably because few of us know very much about Bluetooth. Because Bluetooth technology is not well understood by the vast majority of us it is felt necessary to describe enough of the features of networking software now in general usage for a person to be able to make intelligent decisions when attempting to implement Bluetooth technology. The basic idea of this chapter is to attempt to describe the present networking software that Bluetooth technology is expected to replace in whole or in part. Upon this basic foundation it is hoped that the average person will be able to put Bluetooth technology to good use in his or her daily life.

At times it is expedient to only connect two digital devices together. For example in order to transfer files and graphics between a hand held computer or PDA and a PC computer a serial line connection can be made. As an alternative in the past connections have been made over infrared wireless hardware as well as with a serial line cable. The protocol used in this kind of communication is generally called Object Exchange Protocol (OBEX) This protocol is used in Bluetooth devices as well in in infrared devices. In order to effectively transfer data and graphics using a serial line connection it is usually necessary to synchronize the speeds of the processing units in both digital devices. Early in the development of hand held computers an operating system called Windows CE Services was available from Microsoft and was installed in the hand held computer such as a Personal Digital Assistant (PDA) Microsoft also developed a software program called Windows Active Sync that is used in this kind of connection to effectively transfer files and graphics. Both of these Microsoft programs are still in use today although somewhat more advanced that the earlier versions.

The OBEX protocol is similar to the data and graphics capabilities of the internet protocols which are TCP/IP and HTTP. However the OBEX protocol is designed for smaller computing devices such as a PDA or mobile phone. OBEX uses applications described as push and pull types of applications which results in effective transfers of data and graphics for small devices including Blutooth devices.

A computer on an ethernet wired LAN usually communicates with another computer on the LAN using a so called protocol called TCP/IP. A protocol is a program that is used by computers to format packets of data or graphics in one computer that can be used by the same program by the receiving computer to process the packets. Each packet has a header with instructions on how to use the data and also the data being transmitted from one computer to another.

TCP/IP protocols are used in PC computers, laptops and even some LAN enabled printers and multi-function devices. The internet is run on these protocols as well. Each digital device using TCP/IP protocols has a unique IP address for identification. A typical IP address uses as many as nine digits, separated by periods. For example an IP address of a computer on a typical LAN could have address 192.168.1.1. Connections between two computers using TCP/IP are established by one computer broadcasting the IP address of the other computer on a LAN. The second computer connects to the first computer then acknowledges the connection by replying a synchronous request to the first computer using the IP address of the first computer. The two computers can then send and receive data and graphics files using the TCP/IP protocol to process the data or graphics.

There are two parts to an IP address. Part one identifies the network the digital device is connected to and part two identifies the specific computer on the network. Wired LAN computers are usually interconnected to each other using a switch or hub. To access the internet a router is used that uses two IP addresses that identify two different networks. For example a router could use the IP address of an ISP connected to the internet then use another set of IP addresses that identify a LAN and the individual computers on the LAN. It is the job of the router to establish connections from one LAN computer to another computer on the internet or some completely different network.

An IP address can be fixed for a computer or can be dynamically assigned by a program called dynamic host connection protocol (DHCP). The important point to keep in mind using any protocol is that each computer must use the same identical protocol to be able to connect to another computer or digital device. Different protocols are used for different purposes and can be used in addition to TCP/IP or other protocols. For example the peer-to-peer also called point-to-point protocol (PPP) is used with serial line connections such as for a computer using a serial line telephone cable connection to a modem. TCP/IP protocols are used for ethernet wired LAN connections and for connections of a digital device to the internet.

Many of the protocols in use today and in the future will be briefly described from time to time in this guide when the need arises for understanding of some project. Just continue to keep in mind that two digital devices cannot connect to each other unless both devices use the same identical protocol or protocols. This is very important to keep in mind because TCP/IP connections are usually used in wired LAN connections but may not be used in wireless connections. Also it is important to keep in mind that wired ethernet LAN connections form just one kind of network using TCP/IP protocols. In the wireless realm of LAN or PAN networks connections may be made using other protocols and even resulting in several kinds of networks.

For example wireless connected computers and digital devices use a network called an ad hoc network. An ad hoc network is useful in connections between smart phones and an ATM for example. When two Bluetooth enabled digital devices or computers are brought in range of another Bluetooth enabled device using an ad hoc network both computers can connect automatically wirelessly. This is important in using a Bluetooth or wireless mobile phone to transact business at an ATM that is Bluetooth enabled or wirelessly enabled.

A so called ad hoc network uses the media access control (MAC) address of each digital computer or device that is connected to another. This is completely different from the so called IP address. Each MAC address uses six numbers separated by colons. The first three numbers identify the manufacturer of the ethernet card or digital device and the last three numbers identify the specific computer or digital device. An example is 00:00:0C:00:09:01. The first three numbers identify the manufacturer of the device which in this case is Cisco Systems. The last three numbers identify the specific computer or device.

After the two digital computers or devices are connected in the ad hoc network using the MAC of each device control of data or graphics transfer is switched to another protocol which could be PPP, TCP/IP or whatever the occasion for the protocol arises.

In addition to the protocols used by typical LAN connections and connections to the internet other software is needed to use the data or graphics that is transferred from one digital device to another. This is accomplished by sharing of files and devices between connected computers or digital devices. For example after a file or folder is shared between two or more computers, each computer can use its software to process the contents of the file or folder just as if the file or folder was created on the local computer and not shared with one or more remote computers.

This is hoped to give the reader a basic understanding of LAN software called protocols which are used in data and graphics transfers. Software used to create shared devices files and folders us then used to allow each connected computer or device to process the contents of the shared material using its own software. This is hoped to permit understanding of many of the numerous Bluetooth protocols and shared material that will be described as time goes on.

The web is another type of network using digital devices. This network runs on the TCP/IP protocols for data transport and the HTTP protocol for data processing of the communication. Another protocol you might run across is called UDP. This protocol is inherently insecure and inaccurate but might be useful for fast database lookups or other tasks over the internet. A person can connect his or her computer or LAN to the internet using a router that is usually supplied by the internet service provider (ISP)

A recent development in this area is the advent of the portable wireless modem that can connect to the internet directly from a laptop or PC computer. The mobile wireless modem also called mobile web is simply connected to the USB connection of a computer and is operated by software installed on the computer usually with a read only memory compact disk. Theoretically other computers on the LAN can connect to the internet through the mobile web modem but this may or may not be possible. In order to connect to the internet then also connect to a LAN network the internet connection must use a router to connect to both the internet and the LAN. A router can be a hardware device or can be implemented with software but is a necessary interface between two different computer networks. If no software router is present then no connection is possible between two different computers and the internet. Also no connection is possible between a wi-fi network and a Bluetooth PAN in this case.

Basic security for small networks is a program called a firewall. A firewall is also useful in basic security for a connection to the internet. Connections to other computers on a network are made using what is called ports or in Windows computers COM ports. A port or COM port is a serial line connection and can be made to operate very fast. For internet connections the TCP/IP transport protocols first connect from an internet computer to a local computer or router through a port. The TCP/IP connection is made through the port then the connection is used to transfer packets of digital signals from one computer to another computer.

Internet connections usually use the TCP/IP transport protocols then on port 80 use the HTTP protocol to process the incoming and outgoing transmissions. For the secured internet connection port 443 is generally used. For this reason a firewall program cannot block ingoing and outgoing connections to ports 80 and 443 if used for internet connections. However, unless the firewall program blocks unused ports connected to the internet or even to a LAN network hackers can gain entry into the computers on the LAN through other ports such as ports 67 and 68. For this reason a firewall program blocks all unused ports to help prevent hackers from gaining entry into the internet connected computer. In addition to opening ports 80 and 443 for internet connections the firewall also has some exceptions to blocking all other available ports. Usually ports 67 and 68 are not blocked because these ports are used for connections that assign IP addresses to computers on a LAN connected through a router the internet. Also be sure to keep in mind that the firewall must open the ports needed by a Bluetooth connection or the connection will be blocked. This is important to know in order to remove trouble from a Bluetooth connection if the firewall is to blame for the trouble.

The SSL or secure internet connection used for banking and the like through port 443 uses encrypted digital communications in addition to other security measures. This encrypted communication generally uses what is called the RSA program to encrypt the transmission. This kind of encryption is generally considered unbreakable code even by Big Brother. In fact one government recently banned using this code for communications because the government was not able to monitor the encrypted communications.

Remember to use the glossary in Appendix A if you need it to fully understand the language of Bluetooth technology.

CHAPTER 4 BLUETOOTH NETWORKS (PAN)

BLUETOOTH CONNECTIONS\

A Bluetooth hardware device is basically a wireless modem that communicates using a small radio transmitter-receiver over a limited range. The device is operated by programs on the sender and the receiver digital devices that the Bluetooth device connects together. It is also possible for the Bluetooth device to do a limited amount of data processing and to use permanent memory for storage of data and graphics. At present this data processing capability and memory storage is very limited but is sure to be expanded considerably in the future. There is no standard size or shape of a Bluetooth device or Bluetooth adapter that is inserted in the USB connection to computers. One small size is called a dongle meaning something small in size.

Bluetooth adapters and Bluetooth enabled hardware devices are presently either one of three classes. Class 1 devices have a usable range of 100 meters or 330 feet. Class 2 devices have a usable range of 20 meters or 66 feet. Class 3 devices have a usable range of 10 meters or 33 feet. In actual usage a Class 3 Bluetooth could have a range of 60 feet so always check the specifications for any Bluetooth device you purchase. These ranges are certain to be extended in the near future.

Bluetooth specifications for services provided and for basic compliance are generated and maintained by an organization called the Special Interest Group (SIG) Members of the SIG include Microsoft, Toshiba, Nokia and Motorola. At present the SIG has set forth specifications for Bluetooth devices version 1.1 and 2.1. Vendors that have Bluetooth devices described as versions 6, 7, 8 or whatever are not standard devices. In order to be able to display the familiar Bluetooth icon that relates to a digital device the manufacturer must have complied with the Bluetooth core specifications and protocols and have at least one Bluetooth service associated with the digital device. To keep up to date on SIG versions and new standards you can navigate to Bluetooth dot com for more information.

This brings up the subject of caveat emptor or let the buyer beware. If you ignore caveat emptor when buying a Bluetooth device or adapter and just buy the device because the manufacturer displays the Bluetooth icon you are likely to be sorry. For example if you want a Bluetooth adapter for a wireless connection for a PAN between your laptop and your PC computer the adapter you buy may not be PAN enabled and you will have to return it to the vendor. This is especially true of mobile phones. Many, many mobile phone manufacturers who advertise that their mobile phones are Bluetooth enabled offer only Bluetooth headset services and Bluetooth hands free auto services. Other services such as graphics transfer, file transfer, bill pay at restaurants are simply not offered the buyer even though the Bluetooth icon is displayed for a device offered by some manufacturer. In other words if a service is not clearly in writing offered by a manufacturer for a Bluetooth device that service is very likely to be NOT offered. In fact some manufacturers even offer

Bluetooth services that are very misleading otherwise. This kind of practice is very misleading to consumers because it leads consumers to believe that Bluetooth devices and adapters have very little usefulness.

So called wireless networks use the same frequency band as do Bluetooth devices namely the 2.4 Ghz band. These so called wi-fi devices use the so called 802.11 standard hardware as set forth by the IEEE for compatibility with other similar devices. Mostly these devices use the 802.11e and the 802.11g standards. On the other hand Bluetooth devices use the so called wireless application protocol (WAP) which is not compatable with 802.11 wireless networking devices. In other words a Bluetooth device cannot at present connect directly to a so called wi-fi LAN. However, a Bluetooth device can connect to each computer on the wi-fi LAN individually. In fact each computer on the LAN can at present connect to up to eight different Bluetooth devices a restriction that is certain to be soon extended. All of these Bluetooth connections are local or NOT provided by an ISP so they are free of any charges.

Remember for a network to network connection a hardware or software router is needed. Thus for a wi-fi network to connect to a PAN Bluetooth network all that is needed is a hardware or software router. This is a marketing problem in that if such a connection is possible there will be little, if any, more need for wi-fi connected networks except for a LAN network to connect to the internet network.

INSTALLING A BLUETOOTH ADAPTER

Many peripheral Bluetooth devices have built in hardware that enables the device to communicate with another Bluetooth enabled device. However, at present computers need to be Bluetooth enabled with a Bluetooth adapter. This adapter is small in size in general and can be called a dongle, meaning something small. In addition to the hardware Bluetooth adapter software must also be installed in the computer to operate the Bluetooth hardware adapter. This operating software is called a driver which can be supplied by the manufacturer of the Bluetooth adapter or by the operating system of the computer. Most computers use the windows operating system in various versions but other computer operating systems exist. After the Bluetooth hardware and software is installed on the base computer the computer can connect to other Bluetooth devices including other Bluetooth computers. With Windows operating systems Bluetooth connections were not given serious considerations prior to Windows XP service pack 2. For Bluetooth connections to computers prior to XP Swervice Pack 2 the manufacturer of the Bluetooth adapter or device needs to supply the Bluetooth operating software called a driver. This is a possibility but can be difficult to fully implement. Many experienced computer technicians are capable of doing this kind of a job.

For one Bluetooth device to connect to another each device must be in the so called discovery mode. For an ad hoc network connection which is automatic as soon as one Bluetooth device comes within the range of the other both Bluetooth devices must have ad hoc networking enabled and both must offer ad hoc networking as a service to other Bluetooth devices. So even if you have ad hoc networking enabled and discovery enabled on your Bluetooth mobile phone and you come within range of a gas pump or ATM you will not connect to the Bluetooth enabled ATM or gas pump unless these devices are also ad hoc networking enabled and offer ad hoc networking as a service.

In the event that you are not interested in the automatic ad hoc network connection with another Bluetooth enabled device it is still necessary for each Bluetooth device to be in discovery mode. When in discovery mode the Bluetooth link managers and security programs take effect so that you will have a reliable and secure connection.

Initially for Bluetooth devices to be able to connect to each other from time to time they must first be paired. After the paring process takes place the paired devices can connect to each other as long as both devices are in range and both turned on and have discovery enabled. Assuming that it is desired to pair two Bluetooth devices one device needs to start the pairing process by requesting that another Bluetooth device be added to its network or that any Bluetooth device within range with discovery turned on be found. The pairing process is basically the same for all Bluetooth devices. The two Bluetooth devices exchange the names of the computers or devices and exchange identical codes used for encryption between the paired devices. After the pairing takes place the two computers can communicate provided each computer at least has identical services. Services include PAN services, ad hoc networking, file transfer, headset service and others. Bluetooth devices identify each other using the MAC address of each device and not using IP addresses as is required over the internet using TCP/IP protocols.

Using the Bluetooth WAP protocol for connections each Bluetooth device is capable of acting as a client or as a server on a client-server connection. In a client-server connection the client sends requests to the server and the server sends a reply to the client. Many internet connections are of this type. For example a person on his or her PC computer connects to the internet through the ISP connection and sends what is called an HTTP request for a web page to the web server which is remotely at the facilities of the ISP. The remote server replies to the local client computer by sending the requested web page to the client which the user can view on his or her PC screen. The WAP protocol effectively uses this kind of connection over small devices such as a small mobile phone or a small Bluetooth adapter.

Another kind of connection between computers is called a distributed computing connection. Instead of using the HTTP protocol PC computers use the shared object application protocol (SOAP) and the web shared data language (WSDL) which is also known as XML. Distributed computing is necessary for many web applications such as encrypted communications over the web. Another example of distributed computing is using a credit card to conduct a purchase over the web. The customer enters the credit card information at the web site of the vendor. The vendor then sends the credit card data and a request for approval to the credit card company. The credit card company processes the data with its computers then may approve the transaction for the vendor and also may charge the credit card account with the purchase. The approval is sent after being computed by the credit card company back to the vendor. The vendor then takes the approval data and processes it at its facility and then informs the buyer that the transaction is approved and the item purchased will be shipped along with other processed data to the customer. This is much more than a simple HTTP request for a web page and a reply sending the requested page to the PC user.

It is interesting to note that the concept of distributed computing has evaded the understanding of authors of Bluetooth specifications from the SIG. These authors identify distributed computing between one client-server computer and another client-server computer but call it mystery computing or use some other name than distributed computing. Not to worry. All this illustrates is that Bluetooth technology is in a world of its own and can be very complex to understand even by computer scientists.

WAP is designed to conduct this kind of distributed computing on a much smaller scale than what generally takes place on the internet. This is how a Bluetooth device can be used to pay bills at the gas pump or to transact business at an ATM wirelessly. Bill paying and processing ATM transactions can only be done by using the distributed computing capabilities of the devices used. This is not only true of secure encryption but also true in the case of complete processing of the data involved by both connected computers.

Another feature of wireless connections for computers is the so called spread frequency or frequency hopping of the radio frequency transmissions. In the past radio frequency transmissions were made on a single carrier wave. A more advanced method of transmission is for the frequency be changed over the entire spectrum of the radio frequency band or over all channels available for Bluetooth communications during the course of the communication. This makes it very difficult if not impossible as a practical matter for a malicious hacker to monitor a Bluetooth radio wave transmission. Both 802.11 wireless devices and Bluetooth devices are capable of frequency hopping or spread spectrum transmissions. Even if a hacker tried to decode the system of frequency changing or hopping using the spread spectrum transmission the hacker would still have to decode the virtually unbreakable coding of the transmission. This makes a Bluetooth device used for credit card payments or at an ATM considerably more secure than any wired system or connection could be. The WAP Bluetooth protocol even uses the so called AX-25 protocol over a much smaller range than that of regular AX-25 communications.

Bluetooth technology is one of the major breakthrough electronic technologies that has been developed and because it is so advanced compared to presently used systems or devices it is usually little or imperfectly understood by the average person or even persons with a technical education and technical experience. However, in order to avoid some or all of the many pitfalls in using Bluetooth devices it is necessary to at least basically understand the principals involved in the usage of the device. Remember not only caveat emptor but Appendix A to this book when involved in a Bluetooth project.

CHAPTER 5 BLUETOOTH DEVICES

INSTALLING AND USING BLUETOOTH DEVICES

The first step in installing a Bluetooth device on a computer is to get a USB extender called a USB hub if you only have one or two USB connections on your computer. Next also get one Bluetooth connection cable extender for each Bluetooth device you install. There are a number of Bluetooth USB extender cables so be sure you get the right one for your application. Just ask the sales person at the electronics store if the cable will work on your application.

Bluetooth devices are based on standardized profiles which are either core Bluetooth profiles or Bluetooth user profiles. The profiles in turn use protocols like the well known TCP/IP protocols to operate. Finally the profiles using the necessary protocols can offer the customer specific services. Basic profiles include the wireless application protocol (WAP) and the object transfer protocol (OBEX) WAP is an internet like transfer program but designed for smaller low bandwidth and is very efficient when used in a Bluetooth device.

The OBEX protocol emulates the RS-232 serial line connection for file and graphics transfer and is used in connection with the RFCOMM protocol for wireless communication. This results in what is called the serial port profile used for file and graphics transfer between Bluetooth devices. The named services using these profiles are file transfer service and graphics transfer service between Bluetooth computers. In order for the file transfer service and the graphics transfer service to operate both Bluetooth enabled computers must have the same general profiles, the same protocols, and the same named services.

Unfortunately it is not presently possible for the customer to sometimes know in advance what services, profiles, and protocols are being offered with a specific Bluetooth device or adapter for a computer. For this reason it is usually a mistake to try to buy a Bluetooth device over the internet unless you know the specific product has been tested for your application and will work. If you buy your Bluetooth device locally it is usually expedient to ask the sales person if it will work then return it if it does not do so. Just because a Bluetooth device or adapter you might buy lists specifications that have the protocols and services you want does not mean the other computer will work properly with it. The bottom line is that both Bluetooth devices to be connected must at the very least have the same identical Bluetooth specific services which presumably will comply with current SIG specifications.

Believe it or not during the preparation of this book a Bluetooth installation failed to have the OBEX profile installed. Another installation did not have the RFCOMM profile installed and working so both installations are so flawed as to be useless.

1. The steps for the initial connection called pairing are basically the same for all connections:

2.   Both devices must be set to discover other Bluetooth devices within range;

3. One device must have add a Bluetooth device activated or must be searching for other Bluetooth devices within range and activated;

4. After the second Bluetooth device is found by the other device an icon or device name will appear on the found device screen;

5. Select that icon or device name for pairing;

6. If a device name does not appear on the found device enter a device name if requested;

7. Select next or ok for the pass code screen or for the screen that asks if you will allow the other device to pair with your device-select ok or next;

8. The screen on the found device will usually show pairing successful or enabled or the like;

9. Confirm the paired connection;

10. You may be prompted to not ask for pairing again. This should be the first and only time that pairing is required for the two devices just paired.

INSTALLING A HEADSET

Probably the most needed application for a Bluetooth device is for a Bluetooth headset to be paired with a Bluetooth mobile phone. Additionally a Bluetooth enabled headset can also be paired with a computer for voice communications in lieu of a wired microphone or can be installed as part of a so called hands free auto system.

What follows is a real life example of installing a Bluetooth headset to a Bluetooth enabled mobile phone. The Bluetooth headset is an Emerson model EM228WM and the mobile phone is a Cricket model MSGM8 mobile phone.

First a check was made in the mobile phone user guide to see if a Bluetooth headset could be paired with it. Probably all mobile phones that use the Bluetooth icon can be paired with a Bluetooth headset but this step is necessary for any Bluetooth application. Sure enough the user guide stated that a Bluetooth headset could be paired with the phone within a range of 30 feet. The user guide also gave instructions for information about the Bluetooth services provided by the phone. The only services provided are headset service and hands free audio service. However this is adequate for pairing with the headset, Other information was also provided by the mobile phone, the phone name; the class which is mobile; and the MAC address which could be necessary for an application.

Directions are given in the phone user guide for setting up the phone to operate its Bluetooth services. First the phone is set to on for Bluetooth and visibility is set to show to all so other Bluetooth devices can see the phone icon after pairing takes place. The phone indicated that the phone would only be temporarily visible to other Bluetooth devices which could mean that this setting would have to be manually changed for a headset to pair and be used with the phone. This later proved to be not necessary, however. With these setting made the headset now will have to the discoverable and operating for pairing to take place.

The headset came with a user guide with directions for connecting to a mobile phone.

First the headset battery had to be charged with the included charger;

The instructions gave the usable range of the headset as three feet so the device had to be placed near the mobile phone for pairing to take place;

1. Directions are given to first set the headset into its pairing mode manually;

2. Next step is to set the mobile phone to search for Bluetooth devices;

3. When found the name of the headset EM228 is displayed and selected by the mobile phone.

4. A pass code is requested and using he headset user guide 0000 was entered and the mobile phone was requested to pair.

5. Finally the screen on the mobile phone displayed that pairing is completed by listing the name of the headset on the screen. The phone was then connected and ready to be used by the headset. After a test was made of the connection power was turned off at the headset and the headset was disconnected by the mobile phone. The headset can answer a call by simply pressing the power on button if the headset is left off to save battery power. The call can also be disconnected by the headset.

This illustrates how Bluetooth devices can be connected and used and also that before buying any

Bluetooth device or devices you must first read the manufacturer's specifications to be sure that both devices offer the same services. According to the literature of some manufactures one of their Bluetooth devices may be configured to use a Bluetooth profile such as the headset profile but that is not necessarily offering a Bluetooth service called headset service.

Bluetooth headsets can also be paired with a computer for use instead of a wired microphone but this is one of those procedures that require careful planning and usually an electronic e-mail to the device manufacturer to confirm that a pairing can be successfully made. It is not recommended that any pairing like that be done only with locally bought Bluetooth devices with a return policy of the vendor that would let you bring back the device if it does not work properly.

Another Bluetooth application is the so called hands free audio pairing used in automobiles. A hands free audio kit is sold by the Bluetooth vendor that is usually attached to the visor of the car with a microphone that allows connections to one or more Bluetooth mobile phones that offer the hands free audio services. This is yet another Bluetooth application that needs to be thoroughly tested using the applicable user guides for directions before being sure that the application will actually work satisfactorily. With Bluetooth applications using is believing regardless of what marketing literature you may see about a Bluetooth device so generally it is best to buy Bluetooth devices locally from a vendor with a good return policy.

INSTALLING A BLUETOOTH ADAPTER AND SOFTWARE ON A COMPUTER

This application is full of pitfalls and any pairing of this nature requires that the buyer thoroughly test the Bluetooth devices before deciding not to return a device to the vendor. For example sales literature may simply list the Bluetooth icon and claim to be Bluetooth enabled but in practice not be usable. Even if the sales literature claims to be usable this can even be misleading. For example sales literature claiming to be an adapter usable with Linux or Windows ME or Windows 98SE can be very misleading. For example the system requirements for any Bluetooth adapter is likely to be Windows XP or higher with a pentium class processor running a 200 Mhz clock with at least 128 MB of fast memory (RAM) Even if these system requirements are met a Windows 98SE and ME computer can only be used as a Bluetooth client.

A client computer can only make requests of a server computer but cannot reply to the server computer requests. For example a client computer could request a file or graphics transfer from the server computer but could not send a file or graphics to the server nor could it be used to pay restaurant bills or at the ATM. As for a Linux operating system thus far no usable Bluetooth applications have been found up to and including Fedora core version 9. Apple computers only support limited Bluetooth services also.

As for Windows 2000 a Bluetooth installation was successfully made but with limited usability. As for Windows

XP and higher versions Microsoft has its own way of installing a Bluetooth adapter which may or may not be compatible with an installation using software supplied by a third party and not Microsoft.

In order to install the Bluetooth adapter on a Windows 2000 laptop computer the adapter was inserted into the USB port on the laptop. The Windows operating system detected the adapter and a search was made by windows for the operating program for the adapter also called the adapter driver. Windows could not find a driver program for the adapter so a request was made to Microsoft to supply the necessary driver downloaded from the Microsoft drivers database for the the adapter. It was not possible to obtain the necessary driver program for the adapter from Microsoft. The bottom line is that it was not possible to install the Bluetooth adapter on the Windows 2000 laptop unless the manufacturer of the Bluetooth device would supply the operating program or driver for the hardware adapter.

A trip to the electronics store located a Bluetooth USB version 2.1 adapter made by IOGEAR. IOGEAR supplied very good documentation that could be used to install this version of a Bluetooth adapter to the laptop computer. The manufacturer instructions were followed closely.

First, the software supplied by the manufacturer on compact disk was installed on the computer. This resulted in a series of installation menus that appeared on the screen of the computer. The installation menus were followed closely step-by-step until the software installation was finished.

Following the manufacturers documentation the Bluetooth icon that appeared on the system tray was clicked which started the initial configuration process. A series of windows were shown on the screen and were followed one by one. A name was entered for the computer, Using the name supplied by the software. Another name could be entered instead, however this was not the case. Next the services offered by the Bluetooth adapter and install software were selected. For purposes of development all of the services offered were selected for usage. In actual practice this may or may not be a good idea. After installing all of the manufacturers supplied software and convicting the Bluetooth adapter, the installation was completed.

After completing the installation the Bluetooth icon on the system tray was clicked and several options appeared on a menu.

The first option shown is explore my Bluetooth places. This opens the Bluetooth places folder. The Bluetooth places folder contains many useful Bluetooth files. The first item shown in this folder is the Bluetooth exchange folder. This folder is used to hold files that are to be transferred to another Bluetooth device. The actual location of this folder is in the my documents folder not the Bluetooth places folder.

The second option shown is add a Bluetooth device. Self explanatory.

The next option is Bluetooth configuration for initial setup of the Bluetooth adapter.

The next option is the quick connect option used to quickly connect to a Bluetooth service that is also active on the other computer. A menu appears on the screen with numerous services with which to connect and use. These quick connect options are as follows:

Send business card

Bluetooth serial Port

Networking access

Dial up networking

File transfer – Find devices

PIM synchronization

Hands free audio

Stereo audio

Audio Gateway

The Bluetooth places folder contains many useful Bluetooth files. Clicking on this option changes the screen to the Bluetooth places folder. The first item shown in this folder is the Bluetooth exchange folder. This folder is used to hold files that are to be transferred to another Bluetooth device. The actual location of this folder is in the my documents folder not the Bluetooth places folder.

The next option shown is my device. This folder has a list of services offered by the Bluetooth hardware and software in this computer. In addition to a simple list the folder has a brief description of the usage of each of the listed services. This list is as follows:

My audio Gateway. This service allows any Bluetooth device such as a headset to replace this computer microphone and speakers.

My Bluetooth serial Port. This service allows Bluetooth devices to connect to this computer via a wireless serial port. Bluetooth serial Port, COM five.

My dial up networking. This service allows remote Bluetooth devices to connect to a remote computers such as an ISP by using a modem attached to this computer.

My network access. This service allows remote Bluetooth devices to share this computer on a network connection which may provide Internet access. Or the service may allow remote Bluetooth devices to connect to this computer for running a private network. A network connection of this computer over a wi-fi network may not be shared with this service. However, each computer on a wi-fi network can be paired with one Bluetooth device.

My file transfer. This service allows remote Bluetooth devices to perform operations on a specific folder on this computer and on that folder's sub-folders and files. Bluetooth exchange folder in my documents folder.

My P. I. M. item transfer. This service allows the exchange of business cards. Also it allows personal information manager(P. I. M.) items such as calendar items, contacts, notes and messages from remote Bluetooth devices.

My head set. This service allows this computer microphone and speaker to become a headset for a mobile phone.

My P. I. M. synchronization service uses a menu that depends on the specific device for which it is desired to synchronize data. Unless a person desires to synchronize P. I. M. data this service is not relevant.

My Bluetooth imaging. This service allows remote Bluetooth devices such as a camera to copy images to this computer. Also the service can be used to transfer any graphics file to another Bluetooth device.

My printer. This service allows remote Bluetooth devices to use a printer that is physically attached to this computer. In order for the printer to work on this computer it is necessary for the computer to have installed the printer operating program also call the printer driver. In order for a remote Bluetooth device to use this printer it is necessary also for the remote device to have installed on that computer the printer driver.

Next is entire Bluetooth neighborhood. This is a display of all Bluetooth devices paired to this computer.

Finally Bluetooth device folder was displayed.

The third major folder for Bluetooth usage or information used to configure a Bluetooth service is found in the control panel. This folder in the control panel is network and dial-up connections. This folder showers connections to the computer. Connections shown are:

Dial up connection;

Bluetooth connection (ethernet);

After installing the Bluetooth adapter and the Bluetooth operating software supplied by the manufacturer on the laptop computer another identical installation was preformed on a PC computer. The laptop computer uses the Windows 2000 operating system and the personal computer uses the Windows XP operating system. At this time both computers have the same identical Bluetooth icon on the system tray in the same identical pop-up menu when the Bluetooth icon is clicked. The same identical information is also shown in the my Bluetooth places folder on both computers

After the Bluetooth adapters were installed on both the laptop and PC computer some of the services installed would need to be configured and used. Most of the services could not be used at this time. Two services installed that would be useful at this time are the file transfer service and the image transfer service.

To use the file transfer service first the service would first have to be configured.

1. From the My Bluetooth places folder found in the control panel the My file transfer folder is selected.

2. Right click properties and set up the Bluetooth exchange folder location to the My documents folder.

3. Next a file folder was copied to the Bluetooth exchange folder in My documents.

4. The Bluetooth icon on the system tray was clicked and the quick connect option is selected.

5. The file transfer choice was selected and find devices was enabled by clicking.

6. The Bluetooth device enabled in the XP computer was found and its icon clicked.

7. From the My Bluetooth places folder the my device folder was selected and opened.

The file transfer was a success.

For the image transfer service an image file can be dragged and dropped to the image transfer icon on one Bluetooth device and the image file could then be opened on the other Bluetooth device connected to the first Bluetooth device. Another successful transfer.

For connecting the laptop computer to the Internet via the PC computer, Internet sharing for this computer first needs to be enabled on the PC. From the laptop the Bluetooth icon is right clicked on the system tray. Find a Bluetooth device was selected. After the PC computer is found by the laptop the dial up connection is right clicked and the username, password, and number to dial is entered. However, it was not possible to actually connect the laptop computer through the PC computer. This could be due to any number of reasons so no more time was spent on this project. Each computer continued to connect to the Internet individually.

An attempt was made to form a personal area network (PAN) by clicking on the networking icon on the PC computer and then attempting to attach a laptop computer to the PC. This also was not successful because a personal area network icon was not displayed on the network connections screen of the XP computer. So this project also was abandoned.

Please keep in mind that this information is for the purpose of illustrating how a real life Bluetooth connection is configured then made and then used. In order for a person to actually connect to Bluetooth computers or to Bluetooth digital devices of any kind the specific documentation for this project would have to be closely followed. Even if the specific documentation for a Bluetooth project is closely followed it is still possible that the connection may not be possible or that the project would result in a situation that is not usable. This could be due to any number of reasons including the manufacturer changing devices or software and not updating documentation.

This completes a description of the laptop computer using Windows 2000 connected to the PC computer using windows XP. Although a personal area network was not established connections that enable files to transfer between computers and graphics to be exchanged from one computer to another eliminates any need for wiring between the two computers. This description also illustrates that sometimes it is not possible to complete a project on the first attempt using digital technology. Sometimes to complete the project in all aspects several weeks or even months might lapse.

Only several of the numerous services were tested in the Windows 2000 to Windows XP connection. File and graphics transfer services were successfully installed and used. PAN and ISP connections were not successfully installed. Other offered services were not tested.

Microsoft did not seriously install Bluetooth operating software using the Windows operating system until XP service pack 2. For this reason to install Bluetooth services on the Windows 2000 operating system the manufacturer software had to be used because no Windows software was in place on the Windows 2000 laptop. The manufacturer software to operate the Bluetooth adapters for the laptop and the XP operating PC computers was installed on both computers as described above. Another installation could have been for the manufacturer software to be installed on the Windows 2000 laptop and then the Windows software could have been used to operate the adapter on the Windows XP PC computer. This configuration was not, however, tested. The next installation tested is the Windows XP laptop to the Windows XP PC computer.

One possible application for a Bluetooth project emerged from this test. This would be a graphics or picture transfer between two Bluetooth devices.. For example pictures on a laptop computer could be transferred via a Bluetooth connection to a PC computer with an attached color printer. The graphics or pictures could then be printed via this connection. Another application would be for a mobile phone with a camera that had Bluetooth picture transfer enabled could transfer the pictures to a PC computer with an attached printer for printout.

INSTALLING BLUETOOTH ADAPTERS ON XP COMPUTERS

This example illustrates an installation of a Bluetooth Personal Area Network on two Windows XP PC Computers. This installation uses Windows operating programs for the Bluetooth adapters. Both computers have Windows XP Service Pack 3 installed.

The next Bluetooth adapter installed on a computer was installed on a Microsoft XP operating system PC computer. The first step in this installation is to insert the Bluetooth adapter in the USB port of the computer and wait for windows XP to install the necessary operating software which is known as a driver. First Windows XP searched for the needed driver and showed on the screen that the driver was not found. A new screen is shown offering an option to connect to the Microsoft database on the Internet and search for a driver there. This option was selected and the search was conducted for the necessary driver on the Microsoft database connected to the Internet. After a long wait Microsoft displayed on the screen that the driver needed for this adapter was found in the manufacture of the driver is the Broadcom Corp. Windows then it installed the driver for the Bluetooth device that is connected to the computer and that completed the installation.

After the installation a Bluetooth icon appeared on the system tray of the computer. When this icon was clicked by

---

the mouse several options were offered on the screen. These options are as follows:

Add a Bluetooth device

Show Bluetooth devices

Send a file

Receive a file

Join a PAN

Remove Bluetooth icon

Open Bluetooth settings

Clicking on the settings option opened a window called Bluetooth devices to appear on the screen. This window has four choices: options, COM ports, hardware, and device properties.

Under the options choices discovery should be activated; under connections allow connections should be activated and alert when connection is requested should be activated also. For the option In the notification area, this should also be activated.

Under COM ports(serial words) there is an option to add COM ports if needed. This option was not needed or used at this time.

Under the hardware option you can choose to display properties or to troubleshoot the Bluetooth installation.

There is also a device properties choice.

In the network connections folder that is accessed from the control panel the XP operating system installed a Bluetooth personal area network (PAN) icon.

In the control panel of this computer a Bluetooth devices icon was installed. Clicking on this icon opens a window with several options. Under the options TAB a window appears on screen that is identical to the window shown under the options choice that is accessed from the Bluetooth icon on the system tray.

To form a Personal Area Network the Bluetooth icon on the system tray was right clicked and the Join A Personal Area Network option was selected. This opened a series of screens that found the other PC computer, installed a pass code, and formed the PAN. An icon for each of the two PC computers with their names was shown on each PC computer as part of the PAN.

Windows has a system of file and printer sharing over a network. First a printer is designated as being shared then on a remote computer the printer is added to the remote computer that is connected to the network. File sharing over a network is done by designating a folder as a shared folder then accessing the files or folders in the shared folder from the My Network Places folder. In order for file and printer sharing to work properly the Windows operating system must recognize the network being used for the resource sharing. In the case of this example the Windows XP operating system only recognized the hard wired network installed on the computers and did not recognize the Bluetooth PAN for purposes of file and printer sharing. What this boils down to is that on the XP to XP Bluetooth PAN installation the PAN computers were joined to the PAN but the PAN is not usable because the Windows operating system would not recognize it for purposes of file and printer sharing.

Another test was made of the XP to XP installation. This was using send a file and receive a file services offered when the Bluetooth icon on the system tray was right clicked. Both XP computers were turned on with discovery enabled. The file was copied to a folder called Sharing in the My Documents folder of both computers. First receive a file was selected in one XP computer then send a file was selected in the other XP computer. Following a series of screens that showed on the display the file was sent from one XP computer to another successfully.

The results of these tests and examples is that using Windows XP computers or Windows version prior to XP very limited Bluetooth connections are available from one computer to another. Later versions of Windows no doubt will correct this situation and permit Bluetooth PAN networks to form and be usable on PC computers. A PAN of a laptop to a PC computer would be usable if there was full file and printer sharing possible and Microsoft will no doubt offer this option on a Windows version. In the meantime unless the Windows version does in fact offer full file and printer sharing on a Bluetooth PAN then any Bluetooth connection from computer to computer will have very limited usefulness.

BLUETOOTH CONNECTIONS TO PERIPHERAL DEVICES

Manufacturers offer a keyboard and mouse combination and separate keyboard and mouse Bluetooth connections to a PC computer. The manufacturer documentation must be strictly followed for this to be successful.

A mobile phone connection to a PC computer would be useful for bill paying at a restaurant or department store, for buying gas at the pump, for ATM transactions and the like but are now only available at some experimental locations or some specific location. For this kind of connection to be possible a mobile phone would have to offer the specific services needed possibly a usable Bluetooth PAN. Otherwise this kind of connection will not work properly.

Obviously extreme caution is necessary before spending hundreds of dollars to make such a connection only to find that there is some glitch in the process that makes the connection not usable.

There are a number of small hand held computers available that need to be connected to a PC computer or the equivalent for some project to work. Formerly this kind of connection took place with a serial line cable or over infrared connections. For this kind of connection to work properly the small computer and the PC computer must have comparable operating systems and the connection must be synchronized. In the past this was done with a program called Windows CE services for the small hand held computer and Windows Active Sync for the PC computer. Present day connections are more complex and may or may not work properly. Microsoft can generally be expected to make a Windows CE services small computer work properly with a Windows version PC computer. However, third party operating systems on the small computers may not work with a Windows version PC computer.

Microsoft generally can be expected to make its hardware and software work properly and usually does so. However, the exceptions to this rule occur in the realm of Bluetooth. In the world of today any Bluetooth connection except for a Bluetooth headset requires careful planning, following good documentation and always have a backup plan such as being able to return products that do not work properly.

STATUS OF BLUETOOTH ON LINUX

Linux is an operating system basically for networks as opposed to the Windows operating system which is basically for home PC computers. Using Linux computers generally requires that the user be a graduate computer scientist or the equivalent so this section of the book will dispense with using simplified terminology. The status of the current state of Bluetooth development related to the Linux operating system by this author will be briefly described here. Two attempts were made to install a Bluetooth adapter on a Linux system, Fedora core 5 and Fedora core 9. Neither was successful. The results briefly follow.

First an attempt was made to install a Bluetooth adapter on a Fedora core 5 machine. The Linux operating system was installed on the computer. A file named bluetooth was located at /etc/bluetooth. This directory contains several files: hcid.conf, pin, and rfcomm.conf. The hcid.conf file is the configuration file for the hci daemon. The rfcomm.conf file is the configuration file for the rfcomm file. A check of the rpm database found several bluez files: bluez-utils, bluez-pin, and bluez-libs. The only services found that were running are hcid and sapd. There was no documentation found on the computer or elsewhere. It appears that the files located are not logically connected. With no logically connected files and no documentation there is no way that this author could even hope to install a Bluetooth adapter on this computer.

A second attempt to install a Bluetooth adapter was made on a Fedora core 9 machine. After installing the operating system a Bluetooth icon appeared on the upper system tray. After clicking on this icon a menu was shown with four options: preferences, about, send file, and browse device.

The preferences option has several tabs: one which displays the mode of operation, the adapter name, and the bonded devices. The services option shows four available services: network service, serial service, audio service, and input service. The screen showed that all of these services are currently running. The last tab shows an option to authorize incoming requests.

The about option refers to the bluez web site. The browse device option offers a screen to select a device to browse. The send file option shows a screen that can be used to browse to a file to send.

A service called bluetooth was located that is enabled and running. Several bluez files were located in the RPM database. No documentation was located. It is not possible for this author to activate a Bluetooth adapter on this computer.

In summary it appears that as far as a Linux computer is concerned the Bluetooth technology is now in a state of development. No doubt on Red Hat Fedora releases after Fedora core 9 it will be possible to successfully install a

Bluetooth adapter to a Linux operated PC computer. In the meantime Bluetooth appears to be only partially developed for Red Hat Linux Fedora core releases. For other versions of Linux the same situation is believed to exist.

CHAPTER 6 PROJECTS

USER EXPERIENCES

My first experience with Bluetooth devices was to be given a so called Bluetooth dongle, the name for a Bluetooth adapter that inserts into the USB connection of a computer. The adapter was in a box with a CD ROM that had the necessary software along with a specification sheet. The specifications said that the adapter worked with Windows 98 so I thought I would use the adapter on my Windows 98 laptop to transfer files to the newer Windows XP computers and the PC internet connection. So I inserted the CD ROM into my laptop and tried to install the needed software for the Bluetooth dongle. All I got was a message on the screen that my computer operating system would not operate the software or the Bluetooth adapter. I later found out that the Bluetooth device would only work on Windows 98 as a so called client that could send requests to a server PC computer and receive replies from the PC and nothing else. Also I found out that I needed 250 megabytes of fast or RAM memory and I only had 64 megabytes. I had neglected to remember the phrase caveat emptor.

That was only one of my first Bluetooth mistakes. I landed in several more Bluetooth pitfalls because I had decided to learn how to use Bluetooth devices in my patent application business. Later after spending several months testing out Bluetooth situations I decided to write this book.

My next mistake was to buy a bargain cellular phone with a camera attached. The phone was advertised as a Bluetooth enabled camera phone that formerly sold for $140 and now was priced at only $90. It seemed like a bargain. The phone worked fine as a mobile telephone and I thought I could make it work as a camera phone also. So I spent a couple of weeks trying to figure out how to take pictures with the camera phone and then transfer them to the PC computer for printout or storage. The end result of this project was that I had bought a camera phone NOT a camera and phone. The camera phone could take pictures and send them to a telephone address but the recipient could not view them and I could not transfer the pictures to my PC for printing. I learned that the only Bluetooth services available on the camera phone are headset services and hands free auto services NOT file and graphics transfer services that I need. By this time it was too late to exchange the camera phone or return it to the seller so I was stuck with a device that I could not use as I wanted. At least I learned another Bluetooth related lesson.

In real life in order to install and use any new computer system or even new hardware or software a game plan must first be made. If a new digital project is successful in all aspects then all is well. However in real life this is seldom the case. It is usually many times more difficult to install a new digital project than to use it. There are tried and proven steps to take in the event that a digital project does not succeed on the first attempt. To begin with a digital project starts with selection of the needed hardware. In the case of Bluetooth this is usually done in an electronics store or over the Internet. If new hardware is purchased at an electronics store usually the sales person at the store can give you some advice or at least inform you of how to return the purchase is it does not work. In the case of Bluetooth very few sales persons understand anything about Bluetooth technology except for using Bluetooth headsets on a Bluetooth enabled mobile phone. Even so it is always a good idea to a ask the sales person for advice. Sometimes after a purchase any problems can be resolved also by calling the salesperson or the department where the Bluetooth device was purchased on the telephone.

Before any new digital hardware of any kind is installed it is always prudent to carefully read the documentation supplied by the manufacturer of the hardware. After reading the the manufacturer's instructions an attempt can be made to install the hardware or the software. If difficulties arise in the first attempt to install the project and the documentation does not supply the answer to the trouble, it is always best to set the project aside and sleep on it so to speak. The next day another attempt can be made to install the project and often this is all that it takes to be successful. If the hardware was purchased on the Internet and trouble occurs in the installation it is often best to call the manufacturer customer service on the telephone or at least to send an e-mail to the technical department of the manufacturer asking for help. Also on the Windows operating system there is a help and support file installed on the PC computer that can be used to sometimes locate a problem with instructions on how to clear the problem of. The best way to solve any problems with installation and use of a digital device is to call a computer technician or computer repair store from the Yellow Pages. This is also the most expensive solution to any problem. If you are unable to be successful in the installation of a digital device and a computer technician cannot solve the problem then it is time to use the return policy of whoever sold you that device. When hiring a computer technician first be sure to specifically identify exactly what you are trying to accomplish. Then be sure to ask the technician if he or she is experienced in doing this task.

For example, probably all sales persons can be helpful in the selection of a Bluetooth headset to use with a Bluetooth enabled mobile phone. When the author purchased a Bluetooth headset for a Bluetooth enabled mobile phone the first step was to check the mobile phone to see if the problem was Bluetooth enabled for a headset. After this was done a trip was made to the electronics store and a salesperson was asked if a specific Bluetooth headset would work on the specific mobile phone. This specific headset is made by Emerson electric company and happens to be the lowest priced headset offered for sale at the electronics store at a price of $20. The sales person confirmed that the Emerson brand headset would work fine on the specific mobile phone which turned out to be true. After the Emerson headset was purchased it turned out that Emerson also had supplied very good documentation for use of the headset. After the manufacturer documentation was closely followed the installation was successful and worked very well after that.

The real-life experience that caused the author to purchase and install this headset is an advertisement run on television by a local personal injury lawyer. The advertisement showed the lawyer with a mobile phone in his hand and he said that he saw a driver of a car using a device like which he held in his hand while driving. The attorney suggested that if anyone watching this advertisement was injured by a driver of a car using this kind of device which happened to be a mobile phone to please call the lawyers office. It seems to the author that a $20 investment that could at least prevent a traffic ticket and might even prevent a personal injury lawsuit for driving a car without using both hands on the wheel would be a very good decision to make. Another factor is that the author has been told that a device advertised on television that hooks up to the car speakers in the mobile phone for hands-free driving that is not a Bluetooth device does not work properly. Even so over one million of these electronic devices have been sold to people. This and other similar experiences have given the author motivation to write this book.

Another successful experience resulted from having Windows XP install a Bluetooth adapter to a PC computer. The Bluetooth adapter was simply inserted into the USB connection of the PC computer and Windows was instructed to find the necessary operating software for the device which was done. The point made by this illustration is that installing hardware supported by Windows software is usually successful on the first attempt. This is also true when a digital device such as a web-camera is install on a PC computer when the hardware is made by Microsoft and the software program to operate the hardware is also a Microsoft product. The Microsoft windows operating system is extremely large and extremely complex to the point where other software and hardware vendors may not be able to design a hardware operating program which is called the device driver that will work properly on a Windows operating system version. It is always safe to install Bluetooth devices on a Windows operating system PC computer when the operating program or device driver for the Bluetooth device is also a Microsoft product. It is not always safe to use third-party vendor device driver software for hardware to be installed in a Microsoft operating system computer.

For example. The author of this book purchased a Microsoft Live Camera for use in video communications using Windows messenger over the internet to clients and prospective clients of the patent application services of the author. The Live Camera hardware had software to operate the device on a compact disk read only memory which was installed in accordance with Microsoft instructions. A test of the device was successful and all seemed to be well on the Windows XP computer. However, after the computer was shut down then restarted later the device would not work. Windows XP could not connect the hardware to the software supplied by Microsoft and Windows XP could not even locate the device itself in its hardware management system. After each unsuccessful try to re-install the device Microsoft asked that a report be sent to the Microsoft database which was done. Apparently this seemed to raise a red flag at Microsoft. Here was a Microsoft device with Microsoft software not working on a Windows XP computer. All I could do is to try several times to re-install the device none of which trials were successful. However, after putting the project on the back burner for a few days everything seemed to work perfectly somehow. All I could figure out is that Microsoft could not let a Microsoft product on a Microsoft XP computer using a Microsoft device operating software called the driver not work properly. Microsoft must have fixed the problem remotely because I authorized Microsoft to do so if necessary previously.

The author of this book has spent many years installing hardware and software on PC and hand held computers with varying degrees of success. Often using third party software on a Windows computer resulted in failed attempts to make the software work properly for a device of some kind. However, as was just stated previously if Microsoft sells a person a device and its operating software called the driver and a person attempts to install the device on a Microsoft computer invariably Microsoft will make the device work if the system requirements for the device are met by the customer.

The author of this book has a computer science education and experience so from time to time acquaintances ask the author to help with a computer hardware or software problem. An acquaintance once as the author to help with a computer program and when the author visited the location of the computer several devices were connected to this computer with a maze of wires. There were so many wires connected to the PC computer that the author could not tell what was connected to what. In addition to the PC computer there was a printer connection, a router connection, and several other connections as well. The author recommended to this person that he should get a Bluetooth adapter for the PC computer at a cost of $20, connect two laptops to the PC computer with two other $20 Bluetooth adapters, connect a mouse and keyboard to the PC computer using wireless Bluetooth connections and otherwise eliminate all the wires except for the wire connecting the PC to a router which was connected to the Internet a wired connection to the display and the line cord. It was amazing to see how much space was taken up by this maze of wires and how the Bluetooth connections made the entire system much easier to use.

Many Bluetooth devices are connected to a PC computer using a serial line connection. Such devices include Pocket PC, Palm PC, hand held PC, and mobile phones. Before Bluetooth technology was used to connect these devices they were usually connected via an infrared connection. The infrared connection supplied the carrier frequency for file and graphics transfer between the Bluetooth devices. For example a so-called pocket PC possibly 6 or 8 inches long by 4 or 5 inches wide with a complete QWERTY keyboard could be operated by a small operating system call Windows CE services. In order for the Pocket PC to transfer files and graphics to and from the PC computer a program called Windows active sync had to be installed on the PC computer. At present this infrared connection is almost always replaced by a Bluetooth connection using the Pocket PC with Windows CE services operating system installed in a connection to a PC computer using a Windows operating system such as Windows XP. This can involve transferring word processing files, spreadsheet files, and even database files. Even address book and business card files on a Pocket PC can be transferred to the PC computer using this arrangement with Bluetooth technology.

Please keep in mind that here we have a PC using a Windows operating system and a small device in the nature of a computer using another Windows operating system called Windows CE services. In a case like this Windows will make the project work. However you do take a chance when you have a small computer like device with a third-party operating system and you attempt to connect this to a PC computer using a Windows operating system version like Windows XP. Whatever you do, do not assume that this is going to work. Whatever you do before you actually buy a small computer like device that does not have the Windows CE services operating system be sure to check out the advantages offered by the Windows CE operating system. And be sure to double check and be sure that you can make a successful connection to the PC using the Windows operating system software. Two different operating systems of a computer are not comparable because each operating system uses a file system that is different from the file system of the other computer operating system file system.

Another pitfall to avoid with this kind of project is to purchase third-party software that could be used to synchronize a small computer like device to a Windows software operated PC computer. Microsoft Windows has several versions of a program called Windows active sync that can successfully be used to connect the PC computer to the small hand-held device. This Windows program can be downloaded from Microsoft free of charge. Usually the Windows active sync program is included on a compact disc with in the small hand-held computer hardware sold. But this may not necessarily be the case.

Recently in the hometown of the author the television news repeatedly had stories about how high technology criminals could sometimes make over $100,000 in a weekend at the gas pump. The way this is done is first the criminal opens the cover of the gas pump and inserts a very small device called a sniffer into the wires leading to the unit in the pump that processes credit card payment information. The sniffer is a device that records in memory all of the credit card information of a person The criminal could buy a key to open any gas pump in town for $15. At the beginning of the week after the weekend the criminal would remove the sniffer from the gas pump early in the morning when he or she would not be detected and use the data stored in the sniffer to steal the identity of everyone who used that gas pump over the weekend.

A sniffer is a hardware or software device that records and stores in memory any digital information that is sent from one digital computer or device to another. In the case of credit cards the credit card information is sent along wires inside the gas pump to the credit card processing unit in the pump in plain text which can be read by any computer since it is not encrypted. Hardware sniffers are small devices that can be purchased for around $20 at an electronics store or over the internet. The hardware sniffer is easily installed in the gas pump data processing system to record credit/debit card information for identity theft purposes. This criminal activity went on for several weeks before anything was done to even slow it down. The criminals involved are reported to never have been caught.

At present in Europe a so-called smart card that encrypts data used for credit card processing can be used to block any attempt to steal credit card information with a sniffer. However so-called smart cards are still wired devices whereas a Bluetooth enabled connection would be even more secure than a so-called smart card and the wireless.

For this reason one application that is under development for Bluetooth mobile phones is the Bluetooth connection to a gas pump for processing credit card information. This will require extensive new hardware and it will require that Bluetooth mobile phones provide the necessary service before this technology can be implemented on a large scale.

Believe it or not a news story on television showed a security video of a criminal actually installing a mechanical sniffer in a ATM at a local bank. Security personnel will tell you that video surveillance cameras often do not stop crime. However video surveillance cameras will more often than not enable the authorities to capture the criminals. At any rate here is another needed application for a Bluetooth enabled mobile phone that can transfer a files to the bank using an ATM. Again to implement this technology extensively new hardware and software must be installed and mobile phones must be enabled to use this Bluetooth technology. At present there are many millions of mobile phones that are Bluetooth enabled but only for a Bluetooth headset or a Bluetooth hands free kit in a car.

Also under development is Bluetooth technology that can be used to purchase items at a department store and later even pay for these items at the store. At present some restaurants actually offer a payment service to customers with a Bluetooth enabled mobile phone that the customer can use to pay the restaurant bill over the Internet and have the payment charged to the customer's credit card also over the Internet.

Whenever a Bluetooth technology project is undertaken always realize that very few, if any, salespeople or any other people can give you any help with the project because very few people know anything about present Bluetooth technology. Hopefully this book will do something to correct this unfortunate situation.

CHAPTER 7 TRENDS

MEGA TRENDS

The basic mega trend in wireless networking technology which includes Bluetooth is miniturization of components. In the world of today transistors which act as signal switches or as amplifiers are becoming smaller and smaller. Thousands of transistor like components of an electronic device can be integrated on one small silicon based wafer or chip. Even other components such as resisters, capacitors, and coils can be put on very small printed circuit boards or even silicon based chips. Not all electronic parts can be significantly reduced in size but the trend is to reduce the size of any electronic device. Bluetooth technology is an example of this trend. For home use of many so called LAN computer and device networks Bluetooth technology in general makes any previous LAN configurations obsolete in part or completely obsolete.

Another technology that makes electronic parts obsolete is the so called removable storage devices based on iron oxide permanent memory storage devices that essentially obsolete so called disk drives for LAN home networks or even larger networks. These removable storage devices are faster, less expensive, and more reliable that present day so called disk storage devices or magnetic media storage devices. Between Bluetooth technology and iron oxide removable storage media the electronics realm of today is going to undergo a major transformation in the immediate future. This unfortunately will mean many present day electronics companies will either go out of business or will have to enter into merger agreements with other companies.

A third technology that is slowly progressing is the so called voice recognition technology. At present there is a place for this technology even though it is still full of so called bugs and not useful as a stand alone system for dictation or otherwise. It seems that present day voice recognition software programs work best with few choices of sounds to recognize. For example a telephone connected system might ask the person on the line to say yes or no. The voice recognition program has a good chance to recognize which word was spoken. However, for long phrases or sentences this technology is not yet very useful. Nevertheless this technology is very useful in connection with Bluetooth devices because the user can be given only a very few choices for voice commands.

About 40 years ago Intel developed a single silicon based chip that held a complete computer processing unit similar to the big mainframe IBM model 360 that had revolutionized the computing industry. Intel called this chip the Intel 4004 CPU. Two such chips could be connected in series to hold 8 bit computer words and operate word processing programs, spread sheets, the internet, electronic e-mail and other applications. Later Intel released the so called 8008 chip which enabled computer manufacturers to produce very sophisticated PC home and business computing equipment.

What is now Microsoft Windows began as an operating system called DOS which was only 64k bytes in size at the core. DOS had several modules such as copydisk, copy, and the like which was adequate to operate a PC computer with an elementary word processor called Word, an elementary spreadsheet called Calc and electronic e-mail using the first version of the floppy disk which was only 252k bytes in size. Two such floppy disks plus 64k bytes of fast memory also called RAM became enough to operate the first PC computers for home and office use. This is all that people needed at the time and really almost all that people need today even though PC computers have become considerably more complex.

The PC computer of today generally has far more computing power than any individual on a home LAN would need or could use. At present the home PC computer is useful for connecting to the Internet and also for connecting to some peripheral devices such as a printer. For other jobs such as a word processor or a spreadsheet the small laptop computer is more than adequate for most jobs that people need to accomplish in their home. For a home network of one PC computer connected to the Internet and several small laptop computers used for schoolwork or business notes and word processing after leaving the job and going home a Bluetooth PAN network is all that would be needed to connect the devices to each other. Another consideration is that a Bluetooth computer adapter costs only about $20 per computer so that four computers in the home could be connected in a Bluetooth PAN network for a total of $80.

In the world of today there are countless applications for using a small hand held computer or a laptop to operate a larger computer or larger digital device after being connected to that larger device. This is the situation where Bluetooth technology is very useful and cost effective. In the home a Bluetooth headset could be used to turn home appliances off and on. When a person is on the move around and about the home the small Bluetooth headset could be used with voice commands to operate a smart phone for communications. Small hand held computers can be very useful at school to enter notes, schedules and other data then using Bluetooth technology this digital data can be transferred to the PC computer for permanent storage. Home networks that consist of several laptop or handheld computers and one PC computer connected to the Internet can profit from the use of Bluetooth technology. One big advantage to Bluetooth connections is that one computer can now connect to more than one other computer or digital device. At present one computer can use Bluetooth technology to connect to up to eight of their computers or digital devices without the use of a hub, switch or router. This limitation of eight devices is expected to be shortly extended.

Bluetooth technology, removable device memory, and voice command technology is simply not well understood by the people that it can benefit. The more that consumers learn about these three breakthrough technologies the more they will be used. These breakthrough technology's are useful to consumers who have very limited knowledge and experience with electronics technology. Other breakthrough technologies such as the transistor and the integrated circuit on a silicon chip are useful to electrical engineers and computer scientists who are trained to use any technology in system design whether the technology is new or well established. What this seems to boil down to at the bottom line is that the use is Bluetooth and related breakthrough technologies is dependent on marketing by the companies involved.

Anywhere that there are digital computers or digital devices of any kind connected by wires Bluetooth and related technologies could be profitably employed. This employment of Bluetooth devices is basically only dependent on the knowledge of the technologies by the consumers or users. Bluetooth devices can even be used on electrical appliances that are not connected with wires to begin with.

The players in the Bluetooth technology realm began with consumers who have little, if any, knowledge of electronics. Consumers are dependent on salespeople, technicians, and marketing related information. This is expected to continuously expand the usage of Bluetooth related technologies in the realm of digital electronics. It is not expected to change the roles of salespeople and electronic stores electronic technicians, and the marketing efforts of manufacturers. The trend today in jobs available for electronic related purposes more and more demand that the job seeker have at least an associate degree in computer science. The bigger the company for more it will be necessary for a job seeker to have appropriate degree levels in the realm of computer science.

In the world of today Bluetooth technology is being used to connect two types of devices, computers and other digital devices. The trend with computers is for the machines to become smaller and smaller for home and personal use. The most popular digital device other than a computer is the so-called smart phone. The trend with the so-called smart phone is to use more and more features for the smart phone to give it more and more computing power. At present any size computer, small or large, can connect to the Internet without using a so-called router also known as a gateway. This is possible by using a mobile modem that connects to the computer through the USB connection. Another Bluetooth enabled computer could connect to the Internet through this computer with the attached mobile modem. For one network to connect to another network there must be either a hardware or software router also known as a gateway. At present there is no known computer that can connect to the Internet with a mobile wireless modem and also connect to a Bluetooth PAN. This means the computer connected to the Internet must have a software router or gateway installed which is not the case for any known computer today.

In addition to being able to connect to the Internet a base computer must be able to connect to a printer or multi-function machine which serves as a printer, facsimile machine, and scanner. Otherwise the base computer would have very limited usefulness. Obviously, the trend in home computers is to use a base computer that can connect to a printer or a multi-function machine, and wirelessly also connect to the Internet. In addition to this base computer the trend is to connect a so-called smart phone to the base computer as well as a printer.

The smart phone is likely to also be connected to a headset for hands free usage of the smart phone. It is believed that this kind of hardware structure would cover virtually all of the possible usages for a smart phone and a computer for anyone at home, at school, or at work. If you think about it this would make obsolete not only wired connections but many hardware devices currently used in home networks as well. As for the so-called disk driver, clearly the removable disk technology makes them obsolete for usage at home or at most jobs. It is also going to make very large amounts of money for companies that are able to see this mega trend and to jump on the bandwagon.

As for the mega trends in software most mega trends involve a so-called operating system. The first PC computer that was put into a case that could be connected to a storage memory device such as a floppy disk is believed to be the Apple computer. Shortly after the advent of the Apple Computer the so-called IBM or PC computer was offered for sale by many manufacturers. The Apple Computer used its own operating system. The PC computer used mainly the so-called DOS operating system which later became the Windows operating system.

The first computers used the so-called 4004 Intel CPU. Two of these had to be joined in series in order to process eight bit words. After Intel began producing the 8080 chip PC computers only used one CPU chip and the Apple computer continued to use two CPU chips connected in series. This produced a much faster Apple Computer when compared to the PC computer. However when the Windows operating system began using icon's like were used in the Apple Computer most computers sold were the PC computers. Also by far the most popular operating system became the Windows operating system. This continues today.

The Windows operating system is a virtual monopoly for operating PC computers. The only other operating system that poses a challenge to this monopoly is the so-called Linux operating system. The mega-trend in PC computers is for this virtual monopoly of the Windows operating system to continue for home and very small businesses PC computers. There is a trend for the Linux operating system to more and more replace Microsoft products with networked computers of any size. The Linux operating system requires professional computer programmers for its operation but is superior in most respects compared to any other operating system for networks including the World Wide Web. This does not mean that Microsoft is going to go out of business but that Microsoft is very likely to lose more and more market share of operating systems for so-called PC computers.

Another very large emerging market for operating systems is for an operating system for the small so-called pocket PC. Microsoft is the pioneer in the marketing of this kind of operating system. It is called Windows CE and used in conjunction with another Microsoft product called Windows active sync when the small hand-held computer is networked with a larger PC computer. Operating systems are relatively complex programs and one operating system is not likely to interact with another operating system. This gives Microsoft a big advantage in connecting Windows CE small computers with the Windows PC computer.

The reason that one operating system is not likely to connect to another operating system is that part of the operating system called the file system. All different operating systems known have completely different file systems so a competitor in a small computer realm using a third-party operating system is believed to have little, if any, chance of connecting to a Windows operating system on a PC computer. Unless the small hand held or pocket PC can connect to a regular PC computer it is relatively useless and likely to fall by the wayside in the competitive world of today.

The so-called smart phones in use today have more and more computing power built into them. However one weakness of the small smart phones is lack of permanent storage media. Also the smart phones generally do not share large amounts of data to print except in the case of printing camera images produced by the smart phone. For these and many other reasons it is desirable for the smart phones to be able to connect to a PC computer. The best way to do this is to use available Bluetooth technology.

The most versatile connection between Bluetooth computing devices is the so-called PAN connection. Recently Microsoft announced a new smart phone operating system. This means Microsoft can effectively use Microsoft operating systems to connect a PC computer to a hand held computer to a smart mobile phone. Third party operating systems may or may not be able to offer this possibility. This does not mean that Microsoft is going to either go out of business or that Microsoft is going to completely take over the personal computer industry. Microsoft does not now manufacture hardware to any extent. Instead Microsoft continues to specialize in operating systems and other software. Just as Coca Cola lost market share to Pepsi Cola for a period of many years the Microsoft operating system is very likely to continue to lose market share to the Linux operating system as time goes by. Microsoft is very likely to make very large gains in the sale of operating systems for hand held computers, smart phones, and small home networks.

The trend in our political economy is for businesses to become bigger and bigger and other businesses to become smaller and smaller or fall by the wayside. In order for a business no matter how large to be able to become bigger and bigger the business must be heavily committed to diversity and to research and development in the field of electronics. Companies of this nature that are likely to become larger and larger would include companies like Microsoft, HP, Sony, Samsung, Oracle, and many others.

APPENDIX A

BLUETOOTH BUZZ WORDS AND PHRASES

802.11g – A standard created by IEEE for secured wireless networking (wi-fi) This standard applies to wi-fi products such as a wi-fi card or base station. All connected products should adhere to this standard not another standard. Uses the 2.4 Ghz band along with Bluetooth devices, mobile phones and some kitchen appliances. Can support wireless transmissions over a distance of several miles using specialized antennas.

2G, 3G, 4G – second generation, third generation, fourth generation devices

AX-25 – This is a long range radio based protocol used to connect computers on a wide area network similar possibly to the world wide web network. The protocol is currently used with amateur radio equipment and the Linux operating system for very long range wireless connections. Used often by foreign agents or spies in the USA. Also can be used on a Bluetooth device for communications.

RJ-45 cable – Used to connect computers on a wired local area network (LAN) for very fast data and graphics transmission speeds.

RS-232 cable – Used to connect digital devices over a serial line connection such as connection of a PDA computer to a PC computer. This cable connection is replaced by Bluetooth using the RFCOMM protocol.

Bluetooth – A digital device that complies with specifications published by the Special Interest Group (SIG). If the device uses the Bluetooth core programs (protocols) and one or more services (programs) such as hands free audio then the manufacturer is allowed to use the Bluetooth logo or trademark for the device. At present the SIG has only published specifications for Bluetooth devices version 2.1. Devices claiming later versions are not standard Bluetooth devices.

Bluez – A collection of programs used in the Linux operating system to operate a Bluetooth adapter for a Linux operating system. Thus far the author has not found a successful deployment of these programs up to and including Red Hat Fedora Core 9.

Bonding – A process where one Bluetooth device connects to another device using secure wireless transmissions. After bonding the devices can use the same services that are installed on both computers or devices. The same as pairing.

Bus – A parallel line connection between two digital devices or computers located in the CPU of a computer.

Classes – Currently Bluetooth devices are Class 1 with a range of about 100 meters or 300 feet; Class 2 with a range of about 20 meters or 60 feet; and Class 3 with a range of about 10 meters or 30 feet. Some manufactures do not strictly adhere to this definition and will offer for sale a device called Class 3 but with a range of 60 feet so always check manufacturer specifications for any Bluetooth device.

Client – A computer or digital device that can request information from a server computer but cannot reply to requests from another client computer.

Client-Server – A computer or Bluetooth device that can request information from another Client-Server Bluetooth device and can also reply to requests from another Bluetooth computer or device.

Clock – A device that is part of a central processing unit (CPU) that opens and closes the circuits within the CPU in a cyclic pattern. For example the clock will enable the CPU memory write operation then close it and later enable the CPU memory read operation as well as all other CPU operations during the complete clock cycle.

DHCP – Dynamic Host Connection Protocol. A program that assigns IP addresses to computers dynamically as opposed to each computer having a static or permanent IP address. Used by most ISP connections to PC computers.

Distributed Computing – A system of Client-Server computers that are connected and which exchange data processed then sent to one computer then processed by another computer which sends the processed data back to the first computer for further usage on that computer. More involved that a client server connection where the client sends requests to the server and the server replies to the request made by the client computer.

DNS – Domain name service. Used to translate domain names such as www dot site dot com to an IP address that is sent over the internet. After translation of the domain name to an IP address the DNS locates the computer using the IP address on the internet.

Dongle – A term for a small device. Used to label a Bluetooth adapter of a small size used to connect to a computer at the computer USB port.

Driver – An operating program for a device such as a printer operated by its printer driver program.

Ethernet – A program used to transmit data and graphics over a computer network.

Firewall – A system that permits access to certain ports or connections to a computer then blocks access to all other ports or connections. Used mostly on internet connections.

Frequency – An electronic signal usually expressed in Hertz or frequency per second. Used as a carrier wave mostly with much lower frequency signals modulated onto the carrier wave frequency.

Frequency Hopping – A system of communications where the carrier wave is changed dynamically during the transmission. Used instead of communication over just one carrier wave frequency for a much more secure transmission.

Frequency Spectrum – A band of frequencies used for transmissions of specific devices.

Hands Free Audio – A Bluetooth service used to connect a mobile phone or computer to a Bluetooth enabled headset.

HTTP – Hyper text transmission protocol. A programming language used to transmit data and graphics over the internet for use by word processors or other programs on computers.

Hub – A hardware or software device that connects computers or digital devices to a network so that each connected computer or device can be accessed by another connected computer using appropriate protocols (programs)

IEEE – The international of Electrical and Electronics Engineers. An organization that sets standards for electrical or electronic equipment so that standardized devices can work together.

IP Address – A four digit number separated by periods such as 192.168.1.1. The first three numbers identify the computer or device network and the last number identifies the computer or device on the network.

ISP – Internet Service Provider. Used to connect a PC computer to the internet.

LAN – Local Area Network. A small digital device network.

Layer – A part of a stack of programs used in Bluetooth technology. Used by Bluetooth chip designers.

Linux – An operating system for large or small computers used mostly for networks like the internet or smaller networks. Similar to the Windows operating system which is mostly used by stand alone computers or small networks.

MAC – Media Access Control identifier for a digital device. Six numbers separated by colons. The first three numbers identify the device manufacturer and the last three numbers identify the device itself. Used to identify

---

computers and devices on wireless or Bluetooth networks instead of using IP addresses which are used mostly on the internet.

Mobile Internet – An internet set up for small device access such as mobile phones as opposed to the world wide web internet.

Modem – A device used to transmit and receive serial line digital communications.

Mystery computing – A term used by persons who do not understand the concept of distributed computing.

.NET – A system of distributed computing used by Microsoft instead of the SOAP and WSDL protocols used by other distributed computing systems.

OBEX – Object exchange protocol. Used formerly for infrared data and graphics transmissions between small hand held computers and PC computers. Now used by Bluetooth technology for serial line transmissions.

Operating System – Software that operates computer memory, input/output devices, application programs, and file systems. Each operating system has its own file system and is not comparable with another operating system using a different file system unless a bridging program is used to connect the two operating systems.

Pairing – Same as bonding. Each paired Bluetooth device can use the same services on both computers such as the hands free audio service.

PAN – Personal Area Network. A Bluetooth network where each networked device can connect to up to eight other devices.

PDA – Personal Digital Assistant. A hand held computer that can be carried around then connected to a PC computer to transfer data and graphics to the PC computer.

PID – Personal Identification Directory. For example an address book or the like.

Profiles – Bluetooth specifications for applications or services offered by a Bluetooth computer or device.

Protocols – Programs used for different communication purposes. Such purposes include data transport, data applications, and data storage.

RFCOMM – A Bluetooth protocol used for replacement of serial line cable connections.

Router – A device or software that connects one network to another network not just one computer to another computer.

Server – A computer that replies to requests from a client computer.

Service – A Bluetooth operating program for a specific usage such as hands free audio service. Connected Bluetooth devices must all have the same service enabled for usage of each connected computer.

Sniffer – A hardware or software device used to record key strokes entered into a computer before the key strokes are encrypted or before the key strokes reach the computer CPU.

SOAP – Shared Object Application Protocol. A program used in distributed computing that connects two or more computers each of which process data then send it to another computer for additional processing. More than just a client-server system.

Spectrum – A band of frequencies used in radio transmissions.

Spread Spectrum – A system of communication that uses more than one transmission frequency of a band or uses more than one channel at a time for transmissions. This results in a more secure transmission than a transmission using only one frequency.

SSL – Secure Sockets Layer. A secure computer connection using encryption of the data being transmitted.

Stack – A structure of several Bluetooth protocols used for a transmission. One protocol used for transport and another for data transmission for example.

Switch – A device connecting computers on a LAN where all computers are connected to the switch and connected to each other using IP addressing.

TCP/IP – Transmission Control Protocol/Internet Protocol. Programs used to transmit data and graphics over the internet and wired networks.

UDP – User Datagram Protocol. A program used to transmit data over a network. Not reliable or very accurate but useful for fast database lookups on DNS or the like.

WAP – Wireless Application Protocol. Transmission program for small digital devices as opposed to TCP/IP protocols used for internet transmissions of larger digital devices.

WSDL – Web shared digital language. Same as XML. A data structure used to transmit data and graphics in distributed computing.

BLUETOOTH REFERENCE SOURCES

www bluetooth dot com for gadgets and SIG information

Windows Help using Windows XP or greater versions

For Linux systems Fedora 9 or later try Gnome help

Alphabetical Index

www.ingramcontent.com/pod-product-compliance
Lightning Source LLC
Chambersburg PA
CBHW052148070326
40689CB00050B/2521